WHAT WOULD HAPPEN IF...

WE STOPPED USING PLASTICS?

Written by Claudia Martin

Illustrated by Paula Bossio

WORLD BOOK

www.worldbook.com

READING TIPS

This book asks readers to ponder the question *what would happen if we stopped using plastics?* Readers will discover what plastics are made of, why we depend on them, how they harm the environment, and steps we can take to reduce their harm. Use these tips to help readers consider the ripple effects of certain actions and events.

Before Reading

Explain to readers that this book uses cause and effect to show how human activity can affect the environment, living things, and our planet's resources. Cause and effect can help us think about why things happen the way they do. It can also help us think about what might happen in the future because of our actions and choices today. Encourage readers to be on the lookout for examples of a cause and effect structure as they explore what would happen if we stopped using plastics.

During Reading

Discuss with readers how some actions and events have multiple causes and others have multiple effects. Explain that it can be tricky to keep all the if/then scenarios straight in our minds, so it can be helpful to create a visual guide. Encourage readers to draw and add notes to their own cause and effect maps like those found on pages 22-23, 28-29, and 38-39.

After Reading

After finishing the book, discuss with readers how their understandings and opinions of plastics, their usefulness, and their impact on our environment have changed. Additionally, you can have readers respond to the comprehension questions included on page 46 and complete the Chain of Events activity on page 47 to further extend the learning.

Visit **www.worldbook.com/resources** for additional, free educational materials.

There is a glossary of terms on pages 44–45. Terms defined in the glossary are in boldface type that **looks like this** on their first appearance on any spread (two facing pages).

Contents

No more plastics?

Look around you! Can you see any **plastic** items? From balls to water bottles, bags to rain boots, plastics are everywhere! That's because plastics are super-useful. So why would we want to stop using plastics? And what would happen if we did?

So what are plastics? "Plastic" means "easily shaped." When we talk about "plastics," we're usually talking about factory-made materials that, when hot and soft, are easy to shape into many different forms, from buckets to pipes to threads.

There are many types of plastics, most with long names like polyethylene and **polyvinyl chloride**—which is why we usually just call them "plastics"! Most of the plastics in your home are made from **fossil fuels**: oil, natural gas, and coal.

Around 90 percent of toys are wholly or partly made from plastic.

I'm not made of plastic!

Oh, yes you are!

What's the problem with plastics? Plastics are very strong, so they're hard to dispose of when we've finished using them. In fact, plastics can hang around for hundreds of years, creating litter and pollution! So would it be a good idea to stop using plastics? Read on to find out!

DID YOU KNOW?

Up to 50 percent of plastics are used once, then thrown away.

Every year, the world makes about 110 lb (50 kg) of plastic per person.

Over 10.5 billion tons (9.5 billion metric tons) of plastic has been made so far.

The world produces 440 million tons (400 million metric tons) of plastic waste every year.

Around 40 percent of plastic is made for packaging, from bags to bottles.

THINK ABOUT IT!

Can you name three objects you use every day that are made from plastics?

I can't think of any plastic objects ...

Plastic fantastic!

In this chapter, we'll find out more about **plastics** and how we use them. You might be surprised by some of the products that are made of plastic!

Let's take a close-up look at what makes plastics special. Plastics contain **polymers**! Polymers are long, chainlike **molecules**. And what are molecules? They're groups of linked **atoms**, which are the tiniest building blocks of absolutely everything, far too small to be seen by the human eye.

I'm a polymer molecule!

Rubber is a natural polymer. It comes from a sticky liquid made by rubber trees. When the liquid dries, it is an elastic, bouncy, waterproof solid.

Some polymers are natural, but most plastics contain **synthetic** polymers, which means that they're made in a factory. They're made by a process called polymerization, which links atoms into a chain, usually by heating and squeezing.

When plastics are hot, they can be **molded** into any shape. Some can be melted into new shapes again and again.

And what's good about polymers? These long molecules tangle and link together, giving all sorts of useful properties. Polymers make plastics strong. Some plastics are waterproof or elastic, which can make them bouncy. Many plastics are good **insulators** of heat and electricity, which means that they stop heat and electricity from traveling through them.

FUN FACT!

Central Americans started using rubber for bouncy balls in around 1600 B.C.

THINK ABOUT IT!

Consider a plastic product that you own, such as your eyeglasses or a soccer ball. Can you give three words to describe its useful qualities?

Well, I wouldn't say "bouncy."

PLASTIC FANTASTIC!

We're so used to **plastics** that it seems as if they've been around forever! In fact, plastics didn't become popular until the middle of the last century ...

The first plastic, called celluloid, was made in 1862 by British inventor Alexander Parkes, using a **polymer** he created from plants. Celluloid was lightweight, strong, and easy to clean. It gave people an alternative to animal products, such as tortoiseshell and ivory from elephant tusks, which were expensive. That was good for animals, too!

The main problem with celluloid was that it caught fire easily.

This doll was made from celluloid in the early twentieth century.

THINK ABOUT IT!

Why do you think plastics became so popular in the twentieth century?

At first, plastics were very popular with elephants that liked having tusks!

But plastics made from plants—known as **bioplastics**—didn't really catch on. Then, in the twentieth century, people started making plastics from **fossil fuels**—oil, natural gas, and coal—which formed underground from the remains of long-dead plants and animals. Fossil fuels can be turned into polymers more easily and cheaply.

As new drilling and mining techniques made it easier to get fossil fuels, plastics became even easier and cheaper to make! Most people lost interest in bioplastics.

FUN FACT!
Stockings made from nylon—a silklike plastic produced from oil—first went on sale in 1940, selling 64 million pairs in the first year.

Hello! I'm a Belgian scientist named Leo Baekeland. People often call me the "father of the plastics industry" because, in 1907, I invented Bakelite, the first fully **synthetic** plastic. That means that none of its **molecules** are found in nature! I produced my plastic using phenol, which is made from coal. It was the start of something big!

PLASTIC FANTASTIC!

Let's take a look at some of the most commonly used **plastics**. Like 99 percent of plastics made today, all of them are made from **fossil fuels**.

LOW-DENSITY POLYETHYLENE

Properties: It's lightweight and **flexible**.

Common uses: Its most common use is bags, but it's also found in juice boxes and sheaths around electric wires, which make them safe to touch.

HIGH-DENSITY POLYETHYLENE

Properties: This plastic is stronger than low-density polyethylene.

Common uses: It's used for toys, playground equipment, food storage containers, garbage cans, fuel tanks, and street signs.

POLYETHYLENE TEREPHTHALATE (PET)

Properties: PET is lightweight, shatterproof, and a barrier to oxygen, so food and drink don't spoil.

Common uses: It's made into beverage bottles and takeout food containers. When formed into fibers, known as polyester, it's woven into clothing, for example activewear and fleeces.

POLYPROPYLENE

Properties: It's tough, stiff, and doesn't melt until a high temperature.

Common uses: It's used for garden furniture, suitcases, and such car parts as bumpers and steering wheel covers. Because it's easy to make sterile (totally clean), it's also used by doctors for such tasks as stitching wounds.

POLYVINYL CHLORIDE (PVC)

Properties: PVC comes in two strong, waterproof forms—inflexible and flexible.

Common uses: In its inflexible form, it's used for water pipes and construction materials. When made bendier, it's found in flooring, medical tubing, raincoats, and shoe soles.

POLYSTYRENE

Properties: It's made as a fluffy foam or a hard, brittle solid.

Common uses: As a foam, it's used as protection from knocks in packaging, helmets, and car seats, as well as wall and roof insulation so homes don't get too hot or cold. In its hard form, it's found in TV and computer parts.

FUN FACT! Modern space suits are made from a flexible plastic fiber called Kevlar, which is 10 times stronger than steel.

THINK ABOUT IT!

Did you realize how much plastic you use in your daily life? Were you surprised to find out that any products listed on this page are made of plastic?

Are space suits REALLY made from plastic?

Problem plastics

Now that we know why **plastics** are useful, let's look at why they may not be so great! The most common plastics, made from **fossil fuels**, are causing problems for our planet.

Plastics are difficult to get rid of. Products that come from plants and animals, such as food, are easier to dispose of, because they **biodegrade.** This means that they're "eaten" by **bacteria** and other tiny living things. But most tiny living things don't eat plastics. Heat, light, air, and water can break down plastics, but—depending on the plastic type—it takes between 20 and 500 years or even longer.

Like most garbage, plastic waste takes one of four routes: **landfill**, recycling, **incineration**, or becoming litter. We'll look at all four in this chapter, but let's start with landfills. Landfills are holes in the ground where waste is buried.

DID YOU KNOW?
Around 50 percent of the world's plastic waste ends up in landfills.

Blurgh! Stinky!

Modern landfills are lined and covered to keep chemicals from leaking out, but leaks happen in older landfills and some newer ones. Over time, chemicals such as phthalates—which are added to plastics to make them more **flexible**—can soak into groundwater, then seep into rivers and lakes. Fish exposed to phthalates can have problems with egg-laying.

You'd never know it was there, would you?

A full landfill is being covered with plastic sheeting, then soil.

THINK ABOUT IT!

It takes around 450 years for a plastic water bottle, made from PET, to break down. What else do you think will be happening in the world in 450 years?

It's 450 years old!

PROBLEM PLASTICS

Let's look at the second route that waste takes: recycling. Only around 9 percent of the world's **plastic** waste is recycled. Why so little? We all forget to recycle sometimes! And recycling plastic can be difficult, so not everything sent for recycling is recycled successfully.

Plastic can be recycled by shredding it into flakes (pictured), melting it, then forming it into new products.

Most plastics worsen in quality when they're recycled, so most recycled plastics are made into basic products, such as beverage bottles, **polyester** for clothing, garbage bags, and plant pots. To make such strong, long-lasting products as furniture, recycled plastic usually needs to be mixed with new plastic.

Because different plastics melt at different temperatures, plastic items need to be separated into types before recycling. If items are made of mixed-together plastics or have a lid of one plastic joined to a jar of another type, they are often sent to a **landfill**.

In a recycling facility, plastic sorting is done either by smart machines or human sorters, both of which take time and money.

I'm made of PVC!

Some plastics, such as the **low-density polyethylene** used in plastic bags, are more difficult to recycle—so they often aren't! Because this flimsy plastic is used in farming and industry, it needs extra cleaning, which uses lots of electricity.

FUN FACT!

Recycling five PET water bottles creates enough polyester fiber to fill a puffy ski jacket.

NO! Put that in the RECYCLING bin.

THINK ABOUT IT!

Do you always remember to recycle plastic items? What do you think would encourage people to recycle?

PROBLEM PLASTICS

We've come to the third route that **plastic** waste can take: **incineration.** This is when waste is burned! Around 19 percent of plastic waste is incinerated. What's the problem with burning plastic? It's the same problem as with making plastic—it releases a gas called **carbon dioxide.**

In modern incinerators, the heat from burning waste is used to power machines that make electricity. That's good, because we use a lot of electricity!

Because plastics are made from **fossil fuels**, which contain carbon, plastics release carbon dioxide when they're burned. As you've probably heard, putting extra carbon dioxide into the air causes **global warming**, because it traps the sun's heat. Although new incinerators have devices to capture carbon dioxide, older incinerators are still pumping out the gas.

Carbon dioxide is released not just when plastics are burned, but also when they're made in factories. In fact, plastic production and incineration release around 1.1 billion tons (1 billion metric tons) of carbon dioxide every year!

PLEASE put a stop to global warming!

Global warming is leading to more wildfires, such as those that burned 94,000 sq miles (243,000 sq km) in Australia in 2019–2020.

DID YOU KNOW?
Burning 2.2 lb (1 kg) of plastic releases 6.4 lb (2.9 kg) of carbon dioxide into the air.

Hello! I'm an American scientist named Eunice Foote. Way back in 1856, I was the first to warn about the possibility of global warming. I realized that rising levels of carbon dioxide would warm the air and change Earth's climate! It's a shame no one listened ...

PROBLEM PLASTICS

The remainder of the world's **plastic** waste—22 percent of it—takes the fourth route: It ends up as litter or is thrown into an open, unmanaged dump.

Let's take a long, hard look at plastic litter, which is found in cities, in the wild, and in rivers and oceans. A lot of plastic litter is packaging: bottles, bags, wrappers, bottle caps, straws, and stirrers. Cigarette butts—which contain tiny plastic fibers—are the most common form of plastic litter. Yuck!

All these items are lightweight, so they're easily blown—or washed by rain—from uncovered **landfills** and garbage cans. They are also the items that people most carelessly toss in the street, park, or beach.

Around 220 million tons (200 million metric tons) of plastic waste is in our oceans, thrown there, blown there, and carried there by rivers. Once it's in the water, it can take centuries to break down.

Oh no!

Plastic waste can trap or be swallowed by such animals as birds and fish. **Endangered** animals, including elephants, tigers, dolphins, and sea turtles, have been killed by plastic waste.

Luckily, this great crested grebe was not harmed by its horrible experience.

THINK ABOUT IT!

Next time you're in a food store, take a look at all the packaging. How much of it is plastic? Do you think some packaging is too large or even unnecessary?

PROBLEM PLASTICS

Let's look at one last problem with **plastics: microplastics.** These are little specks of plastic. Plastic waste can be rotted, shredded, burned, or broken into microplastic by any of the waste routes we've discussed— **landfills**, recycling, **incineration**, and littering.

What's the big problem with these little plastics? Microplastics are easily blown, floated, and washed—into the air, through the oceans, and into the soil. The specks are so small that they're eaten by animals from fish to earthworms.

Scientists think that there are more than 50 trillion microplastics in the world's oceans.

Microplastics can build up in an animal's stomach, making it feel full so it doesn't eat food. Scientists think chemicals in microplastics can also affect animals—such as oysters—so they lay fewer eggs.

Once microplastics are inside one animal, they can get inside the animal that eats that animal, then the animal that eats that animal. Microplastics can affect the whole **food chain!**

Microplastics among green **algae** are eaten by water fleas.

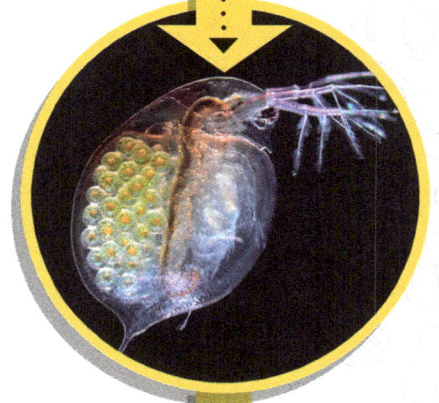

Hi! My name's Chelsea Rochman. I'm an American scientist who is studying how microplastics in lakes and oceans are affecting such animals as crabs, fish, and birds. My work has led me to campaign for better management of plastic waste. But it would be even better if we made less waste in the first place!

Water fleas are eaten by trout.

Trout are eaten by people!

What happens to 100 plastic bottles?

Let's finish our look at the problems with **plastics** by examining a super-common plastic item: bottles. Across the world, 1 million plastic bottles are sold every minute. Let's look at the fate of just 100 of those bottles …

These 100 small plastic bottles are made from **polyethylene terephthalate (PET)**. Around 1 oz (28 g) of **carbon dioxide** gas is **emitted** while producing 1 oz (28 g) of PET, so before our 100 bottles have left the factory, they've released 100 oz (2,800 g) of gas.

Our bottles increase the amount of carbon dioxide in the **atmosphere**, worsening—by just a little bit—the world's problem with **global warming.**

Filled with water, soda, or juice, our bottles are bought by 100 thirsty people, then disposed of …

Around 50 bottles end up in **landfill**. Depending on the landfill, our bottles may leak **microplastics** and such chemicals as phthalates into the environment.

Around 9 bottles are separated from their caps, then sent for recycling. They become useful products, such as running shoes. Depending on the recycling plant, some microplastics may be released into the air.

THINK ABOUT IT!

What do you think is the best way to dispose of plastic waste? Can you think of any problems with this method?

Mmmmm ... incineration?

Around 19 bottles are incinerated, creating electricity for homes and businesses. However, depending on the incinerator, carbon dioxide and microplastics may be released into the air.

Around 22 of our bottles are littered or put in unmanaged dumps. Whole plastic waste and microplastics—broken up by weather and sunlight— are blown and washed away.

Whole plastic waste and microplastics pollute air, land, and water, endangering the health and lives of animals from worms to whales.

Zero plastic

Now you know about the harm that **plastics** are doing! So do you think we should stop using plastics today? In this chapter, we'll find out what would happen if we went zero plastic right now.

First, let's look at the benefits of stopping production of plastics. There would be no new plastic waste! No more plastic would head for **landfills,** incinerators, and recycling plants, which would reduce the number of new landfills, cut pollution, and slow **global warming.**

There would be no new plastic litter thrown in forests, rivers, and oceans. But we would still need to clean up the plastic that is already littering our planet!

DID YOU KNOW?

The top three items found in beach cleanups are food wrappers, cigarette butts, and plastic bottles.

Stopping the production of plastics would also stop plastic-making factories from **emitting carbon dioxide**. Around 4 percent less oil, natural gas, and coal would be mined, because it wouldn't be needed for plastics. The mining and transportation of **fossil fuels** can cause pollution and displace animals—so less mining would be a benefit!

This can't be good for seabirds.

In Thailand, workers mop up an oil spill from a burst pipeline, which was carrying oil from an undersea drilling area.

THINK ABOUT IT!

Would you be glad if the world stopped using plastics today? Read to the end of this chapter, then ask yourself the question again.

No more plastic!

But wait a minute! **Plastics** are used to make all kinds of items. If we stopped making plastics today, we'd need to lose those items or make them out of other materials. Are there any drawbacks to swapping plastic for other existing materials?

Plastic is more lightweight than such alternative materials as glass or many metals. Losing plastic would mean that eyeglasses, phones, and camping equipment could become too heavy to be usable. More fuel would be needed to run heavier cars, trains, and planes. Burning extra oil would release more **carbon dioxide!**

The Boeing 787 Dreamliner is 50 percent plastic, making it so lightweight that it uses 20 to 30 percent less fuel than similar-sized planes.

Plastic packaging keeps food and drinks fresh and clean. Losing plastic packaging would increase food waste, because more food would go bad. Switching to heavier glass or metal packaging would increase fuel use during transportation.

Although most alternatives to plastic—including human-made, plant, and animal materials—are easier to recycle or **biodegrade,** they use our planet's **resources,** too. Less electricity is needed to produce plastic than glass and steel. It takes more water to grow the cotton for a T-shirt than it does to make the **polyester** for a T-shirt. And we really don't want to use animal products, such as ivory or tortoiseshell!

It takes 2,200 gallons (10,000 l) of water to produce 2.2 lb (1 kg) of cotton.

FUN FACT!

Plastic lenses for eyeglasses were introduced in the 1980's. They were less breakable, thinner, and lighter than glass.

THINK ABOUT IT!

Can you think of any ways that the invention of plastics benefited animals? Can you think of any ways it harmed them?

Luckily for me, tortoiseshell-framed glasses are no longer made of tortoiseshell!

What would happen if we stopped making plastic today?

It's unlikely to happen, but let's look at a few of the effects of putting an immediate stop to the production of new **plastics**!

All plastic-making factories close, resulting in a drop in **carbon dioxide** emissions and the mining of **fossil fuels**.

No new plastic waste is created, which reduces plastic litter, shrinks **landfill** use, cuts pollution, and slows **global warming**.

Because recycled plastics are mixed with new plastic to make most plastic products, many common items disappear. Makers of some items switch to alternative materials, making these items heavier, costlier, or more breakable.

Plastic packaging for food and drinks disappears, causing difficulties with storage and transportation.

Can I have a glass of milk, please?

Plastic bank cards and plastic-coated bills disappear.

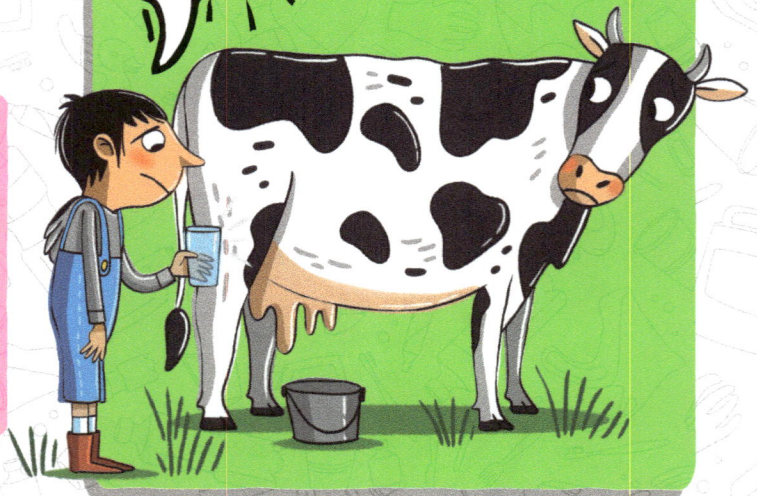

Vehicles become heavier, so more fuel must be burned to power their engines.

Heavier materials are swapped into such electronic equipment as phones and laptops.

Plastic medical equipment disappears, including **syringes**, gloves, and pill packs.

We lose plastic-coated power lines as well as pipes that supply water to homes and fields.

Can you think of any other common products that we would need to live without?

Many cosmetics, clothes, and items of sports equipment cannot be made.

Basketball's no fun without a ball ... and a hoop!

THINK ABOUT IT!

If you had to stop using all your possessions that contain plastic, which item would you miss most?

Steps to stop

As we discovered in the last chapter, we depend so much on **plastic** that we can't just stop making it today! But we can take steps toward two goals: stopping the production of problem plastics and cutting out plastic waste.

First of all, let's look at one area where scientists and manufacturers have already taken steps: **biodegradable** plastics. As we've seen, a big problem with common plastics is that they're not biodegradable, because most **bacteria** and other tiny living things can't consume them.

Amazingly, some newly designed **fossil fuel**-based plastics are biodegradable! For example, polybutylene succinate is biodegraded by particular bacteria and **fungi**. In the future, such biodegradable plastics could be used for packaging, then break down into harmless materials in a **compost** heap.

Polybutylene succinate is made into sheeting that farmers use to keep soil damp and free of weeds.

Sounds like a good idea!

Hello! I'm an Italian scientist named Federica Bertocchini. I want to solve the problem of the tons of plastic waste that are already littering our planet. I'm researching living things that can actually eat plastic, called **plastivores**. I've found waxworms that can break down polyethylene!

I'm hungry for plastic!

THINK ABOUT IT!

Do you think that making new biodegradable plastics from fossil fuels is the only solution we need for the problems caused by plastics? Why or why not?

Are you biodegradable?

Another area where we're taking lots of exciting steps is **bioplastics**. These **plastics** are made from plants or plantlike **algae**.

Just like traditional plastics, bioplastics contain **polymers**. These polymers are either made naturally by living things, such as the starch made by many plants, or they're created in a factory from natural materials.

> I'll put something biodegradable inside these biodegradable bags ...

The big benefit to bioplastics is that they're not made from **fossil fuels**, so they create less pollution. Many bioplastics—such as polylactic acid, made from corn or sugar cane—**biodegrade**, so they're suitable for packaging. Others—such as polyamide 11, made from vegetable oil—have such strong polymers they can be used for long-lasting products, then be recycled.

Polylactic acid is used for dog poop bags, packaging, diapers, tableware, and medical plates and pins, which break down inside the body when their job is done.

Are there any drawbacks to bioplastics? To make large quantities, lots of plants must be grown, which might lead to cutting down trees to clear land. Growing plants requires water, as well as **fertilizers** (to feed the soil) and **pesticides** (to keep them pest-free), which can affect local water sources, plants, and animals.

I can be regrown year after year!

I take millions of years to form!

DID YOU KNOW?

Around 360,000 tons (327,000 metric tons) of bioplastics are produced every year, while 440,000,000 tons (400,000,000 metric tons) of fossil fuel-based plastics are produced.

Hi! I'm a scientist named Jianhua Zhang. I'm making biodegradable bioplastic from seaweeds, which are algae that live in the ocean! We can grow seaweed without taking up land needed for food crops. Even better, seaweed actually soaks up **carbon dioxide** from the air, so it can slow **global warming**!

STEPS TO STOP

A big part of our problem with **plastics** is **single-use plastics**: plastic that is used once for packaging or utensils, then tossed out. But rather than replacing single-use problem plastics with **biodegradable** plastics or **bioplastics**—can't we just use nonplastic materials for these types of jobs? Yes, we can!

BAMBOO

Bamboo is a fast-growing plant with strong stems that can be made into tableware, chopsticks, plates, and toothbrushes. Bamboo is reusable because it's more hard-wearing than paper, but it's lighter than metal or glass. It's also biodegradable.

RICE PAPER

Made from rice flour and water, this paper can be used to wrap food that will be eaten soon after purchase. It's flimsier and less protective than plastic, but it can be eaten after use!

HEMP

This plant is spun into strong fibers, then used to make reusable shopping bags. Hemp is very fast-growing and needs less water and **pesticides** than similar crops, such as cotton.

MUSHROOM ROOTS

Mushroom roots are made into packaging that protects food from knocks, water, and heat. After use, the packaging biodegrades in a **compost** heap in a few weeks.

BEESWAX WRAP

Made by bees, beeswax can be used to coat cotton fabric. The fabric is then washable, moldable, and grippy, so it can replace plastic wrap to keep food fresh in refrigerators and lunch boxes. When it loses its grip after around a year, beeswax wrap is biodegradable.

PASTA

Dried pasta—which is made from wheat and water—is made into drinking straws, which stay strong for around an hour in a cold drink! After use, the pasta is edible or compostable.

PAPER

Paper and cardboard have been used for packaging for centuries! In the twentieth century, much paper packaging was replaced with plastic, metal, and glass, because paper is flimsier and less of a barrier to germs, light, and heat. Paper is becoming popular again, as manufacturers and shoppers change their thinking.

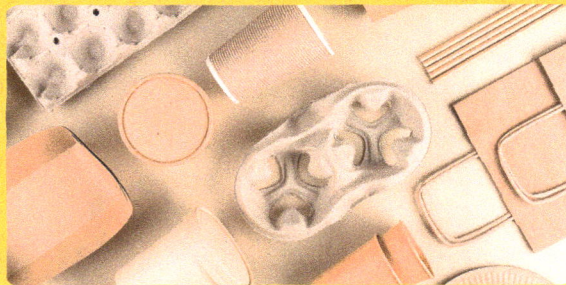

THINK ABOUT IT!

Do you think that a product having bright plastic packaging makes you want to buy it? Would you prefer dull-colored mushroom packaging? Why or why not?

What's wrong with mushroom colored?

Now let's look at ways we can all do our part to solve the **plastic** problem! By reusing and recycling plastic products, we can reduce plastic waste and litter. By avoiding plastic packaging, we can reduce the amount of new plastic being made.

REUSE

Try to reuse plastic shopping bags by taking one with you when you go to the store. Reuse plastic water bottles by refilling them at home. Turn plastic jars, cartons, and yogurt cups into storage containers, plant pots, watering cans, and bird feeders!

RECYCLE

Remember to recycle plastic bottles, food trays, and other plastic packaging. Make sure you first remove lids and other parts that could be made from different types of plastics.

Plastic toys, clothing, and backpacks can be given to friends or donated.

SAY NO!

Try not to buy—or say "no thank you" to—plastic items that are used only once, such as drinking straws and spoons. Some coffeehouses allow you to bring your own reusable cup for them to fill, which saves on plastic cups. If you think a manufacturer of food, toys, or cosmetics uses too much packaging, don't buy their products—and you could write politely to tell them what you think!

FUN FACT!

In 2008, China banned stores from giving free plastic bags to shoppers, resulting in a 50 percent drop in China's plastic bag use.

THINK ABOUT IT!

Can you think of three things you could do to help reduce plastic waste?

How can we tackle the world's problem with plastics?

Let's look at some of the ways that humans can battle—and finally defeat—the problem with **plastics**. We'll focus on two great goals: reducing plastic waste; and, eventually, stopping the production of problem plastics.

Scientists find inexpensive ways to make **biodegradable** plastics from **fossil fuels.** Huge numbers of **plastivores,** such as waxworms and **fungi,** feed on existing plastic waste.

More manufacturers switch to alternatives to fossil fuel-based plastic—from bamboo to **bioplastics.** More shoppers change their attitudes about packaging, so they choose no packaging or reach for dull, brown packaging rather than shiny and bright!

You look so exciting!

Everyone makes a big effort to reuse and recycle the plastic items they already own. Scientists find better ways to recycle plastics, so they can all be reused again and again, without releasing **carbon dioxide** or chemicals into the environment.

Scientists develop bioplastics that are as useful and inexpensive as plastics made from fossil fuels. Efforts are also made to grow plants and **algae** for these plastics in ways that use less land, water, **pesticide,** and **fertilizer.**

38

The world cuts down on plastic waste!

I'll run out of plastic to eat!

THINK ABOUT IT!
Do you think that we'll stop using problem plastics within the next 20 years? Why or why not?

But you can't live without me ... can you?

No new plastics are made from fossil fuels!

Governments make laws that discourage the production of plastics from fossil fuels, while offering support to scientists and businesses that are working on recycling, bioplastics, and other alternative materials.

Conclusion

As you've discovered, we can't—right now—stop using **plastics** made from **fossil fuels**, because we're too dependent on them! Yet, if we all work together, we'll be able to stop using problem plastics one day.

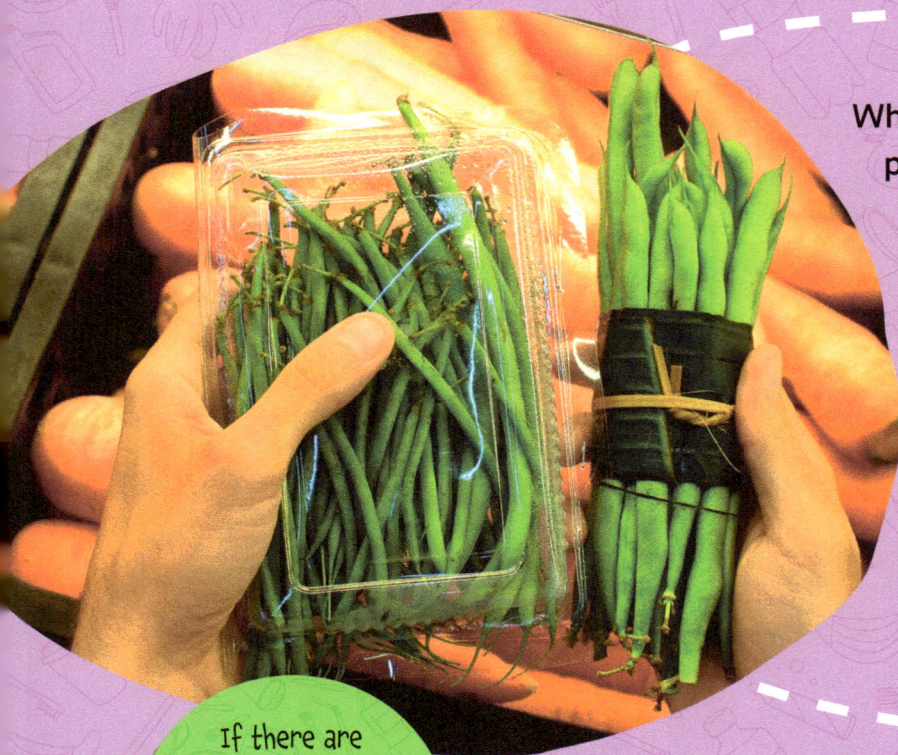

What can you do to battle plastic pollution? You can reuse and recycle the plastics in your home and school. You can remind other people to recycle by drawing posters, telling them what to recycle—and why! You can also choose alternative materials to problem plastics, when they're available and you can afford them.

If there are alternatives, choose products that don't have plastic packaging.

THINK ABOUT IT!

Do you think the most important warriors in the battle against plastic pollution are governments, businesses, or ordinary people?

While we make our personal efforts in this battle, exciting efforts are also being made on a global scale. Around 175 countries are figuring out a treaty (a legally binding agreement) to combat plastic pollution. It aims to wipe out **single-use plastics**, to improve recycling, and to invest in alternative materials.

Ask an adult for help organizing litter cleanups in your local park.

Hello! I'm Tuomas Knowles, a British scientist. Did you know that spider silk is one of the strongest natural materials for its weight? My team have made soy **molecules** link up in the same way as the **polymers** in spider silk. The result is a super-strong **bioplastic** that could replace **single-use plastics!**

Summary

So why are common **plastics** causing a problem? What would happen if we stopped using them today? And since that is unlikely, what steps do we need to take before we can really stop using problem plastics?

Common plastics release **carbon dioxide** during production and **incineration.** Plastic waste—which may take 500 or more years to break down—can leak chemicals and trap or choke animals.

Plastic production stops today!

Gradually, the world tries to cut plastic waste and find alternatives to plastics made from **fossil fuels.**

Many useful or life-saving products—from pipes that deliver clean drinking water to **syringes** that deliver vaccinations—disappear overnight.

Vehicles and other products become heavier, resulting in higher fuel use during transportation—and higher carbon dioxide emissions.

42

Inexpensive, non-polluting bioplastics are produced, with a range of useful properties from **flexible** to heatproof.

The production of plastics made from fossil fuels is phased out. However, sometimes everyone agrees that the benefits of a plastic product—such as a replacement heart for someone with heart disease—outweigh the risk of pollution.

Alternative nonplastic materials—including natural and entirely new humanmade materials—are carefully chosen or designed. The benefits and possible drawbacks (such as land and water use) are considered for each material.

The Earth and its animals and plants are a little safer from pollution and **global warming**.

Existing unwanted plastic items are recycled safely into long-lasting products, from homes to spaceships.

Existing plastic waste is eaten by **plastivores** or collected by an army of plastic-seeking robots.

THINK ABOUT IT!

Do you think that humankind has benefited from the invention of plastics?

43

Glossary

alga (plural: algae)—a plantlike living thing that's usually found in water

atmosphere—the blanket of gases that surrounds Earth, including nitrogen and oxygen, plus smaller amounts of gases such as carbon dioxide

atom—the smallest possible portion of a material

bacterium (plural: bacteria)—a simple living thing, too small to be seen by the human eye without the help of a microscope

biodegrade—to be broken down by tiny living things, such as bacteria

bioplastic—a plastic made from plants or algae

carbon dioxide—a gas, released into the air by burning fossil fuels, that traps heat in Earth's atmosphere

compost—rotting natural materials—including plant and food waste—which can be added to soil to help plants grow

emit—to release or give off

endangered—at risk of dying out

fertilizer—a material that's put on fields to make plants grow well

flexible—able to bend easily without breaking

food chain—a series of living things that feed on each other

fossil fuel—natural gas, oil, or coal, which formed over millions of years from dead animals and plants

fungus (plural: fungi)—a simple living thing that feeds by soaking up rotting living things

global warming—an increase in temperatures on Earth caused by human activities that release such gases as carbon dioxide, which traps the sun's heat

incineration—destroying waste by burning it, while using the heat to power machines that make electricity

44

insulator—a material that stops heat or electricity from escaping

landfill—a pit where waste is buried

low-density polyethylene—a common plastic, made from oil or natural gas, that's often formed into shopping bags

microplastic—a small piece of plastic waste, usually less than 0.2 in (5 mm) long

mold—to shape a material

molecule—a group of linked atoms

pesticide—a substance that's used to kill or harm insects, fungi, or other living things that can damage crops

plastic—a factory-made material—usually produced from fossil fuels—that contains polymers and can be molded when hot

plastivore—a living thing that eats plastic

polyester—a type of polyethylene terephthalate that's formed into fibers which can be made into fabric for clothing, home furnishings, and car safety belts

polyethylene terephthalate (PET)—a common plastic, made from oil and natural gas, that's often formed into beverage bottles and polyester

polymer—a long, chainlike molecule; materials that contain polymers have useful properties, such as being moldable, waterproof, or insulating

polyvinyl chloride (PVC)—a common plastic, made from oil or natural gas, that's often formed into pipes, window frames, flooring, footwear, and wire insulation

resource—a useful material, such as water or coal

single-use plastic—a plastic product—such as packaging, tableware, or drinking straws—that's used only once, for a short time

species—a group of living things that look and behave alike, such as humans or lions

synthetic—containing molecules made in a factory rather than found in nature

syringe—a piece of medical equipment used for sucking liquid out of something or pushing liquid into something

Review and reflect

COMPREHENSION QUESTIONS

Plastic fantastic!

- Why is polymer good for making plastics?
- When did people start making plastics from fossil fuels? What are fossil fuels made from and how are they formed?

Problem plastics

- Why are plastics difficult to get rid of?
- Like most garbage, plastic waste takes one of four routes. What are they and what problems do they cause?

Zero plastic

- What are some of the benefits of stopping the production of plastics?
- What are some of the drawbacks to swapping plastic for other existing materials?

Steps to stop

- We depend too much on plastics to stop making them. But what are some steps we can take toward minimizing their harm to the environment?
- What are some examples of nonplastic materials that can do the job of problem plastics and bioplastics?

Conclusion and summary

- After reading this book and considering what would happen if we stopped using plastics, what is your biggest takeaway? Why?

MAKE A CHAIN OF EVENTS!

Creating a paper chain can help you explore and visualize how cause and effect relationships can be thought of as a sequence of events.

You'll need:
- Pencil
- Scratch paper
- Pens or markers
- Stapler and staples
- Strips of paper (2 colors, if possible)

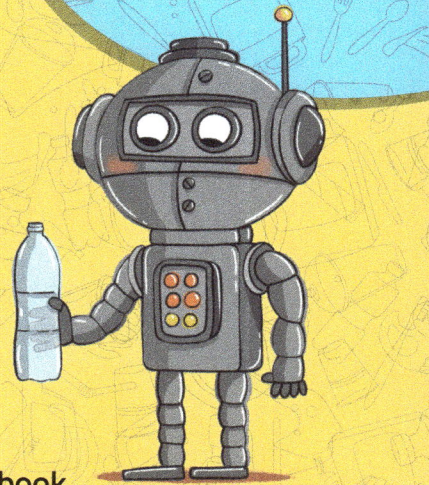

Instructions:

1. **Select a focus:** Choose a specific aspect from the book that caught your attention—it could be how your life would be different if you stopped using all your possessions that contain plastic, or how the invention of plastics has helped or harmed animals.

2. **Brainstorm causes and effects:** On a sheet of scratch paper, brainstorm and list the causes and effects related to your chosen focus. Think critically about the factors that contributed to or resulted from your focus. You can always look back in the text for ideas!

3. **Write on strips:** Write each cause and each effect on its own strip of paper. If you have different colored paper, use one color for the cause strips and the other for the effect strips.

4. **Create the paper chain:** Organize your strips into causes and effects. Start forming a paper chain to show how a cause leads to an effect. Use the stapler to connect the two strips. Continue adding cause and effect strips as links in your chain. When you've finished, you should be able to start at the beginning of your chain and read through each chain link in a logical order.

5. **Linking multiple chains:** If your focus has multiple causes or effects, you can create additional chains and link them together to show how complex cause and effect relationships can be!

Write about it!

Look at the paper chain you created and how the causes link to effects (which in turn link to other causes!). How might breaking a link in the chain impact the overall sequence of events?

World Book, Inc.
180 North LaSalle Street
Suite 900
Chicago, Illinois 60601
USA

For information about other World Book publications, visit our website at www.worldbook.com or call 1-800-WORLDBK (967-5325).

For information about sales to schools and libraries, call 1-800-975-3250 (United States), or 1-800-837-5365 (Canada).

Library of Congress Control Number: 2024941779

What Would Happen If...
ISBN: 978-0-7166-7125-1 (set, hard cover)

We Stopped Using Plastics?
ISBN: 978-0-7166-7126-8 (hard cover)
ISBN: 978-0-7166-7138-1 (e-book)
ISBN: 978-0-7166-7132-9 (soft cover)

Staff

Editorial

Vice President
Tom Evans

Editorial Project Coordinator
Kaile Kilner

Curriculum Designer
Caroline Davidson

Senior Editor
Shawn Brennan

Proofreader
Nathalie Strassheim

Graphics and Design

Senior Visual
Communications Designer
Melanie Bender

Digital Asset Specialist
Rosalia Bledsoe

Written by Claudia Martin
Illustrated by Paula Bossio

Developed with World
Book by White-Thomson
Publishing LTD

Acknowledgments

4-5   © Vorobyeva/Shutterstock; © Alen Kadr, Shutterstock
6-7   © Zyabich/Shutterstock; © Yatra4289/Shutterstock
8-9   © Grizanda/Shutterstock; © INTERFOTO /Alamy Images
10-11   © muratart/Shutterstock; © Evannovostro/Shutterstock; © AR Pictures/Shutterstock; © Gabo_Arts/Shutterstock; © sutadimages/Shutterstock; © ARIMAG/Shutterstock
12-13   © Michael Neelon, Alamy Images; © Dalibor Danilovic, Shutterstock
14-15   © Roman Zaiets, Shutterstock; © ImagineStock/Shutterstock
16-17   © ZUMA Press, Inc./Alamy Images; © Agencja Fotograficzna Caro/Alamy Images
18-19   © Heying HUA/Shutterstock; © Roman Mikhailiuk, Shutterstock
20-21   © Kletr/Shutterstock; © Jacek Chabraszewski, Shutterstock; © Sansoen Saengsakaorat, Shutterstock; © Kichigin/Shutterstock; © Lebendkulturen.de/Shutterstock

22-23   © BearFotos/Shutterstock; © Pixel B/Shutterstock; © lzf/Shutterstock
24-25   © jukurae/Shutterstock; © Davide Angelini, Shutterstock
26-27   © vaalaa/Shutterstock; © CRS PHOTO/Shutterstock; © picturepixx/Shutterstock
30-31   © AG Photo Design/Shutterstock; © Science Photo Library/Alamy Images
32-33   © otsphoto/Shutterstock
34-35   © Elena Milenova, Alamy Images; © Santi S/Shutterstock; © Fevziie/Shutterstock; © Iryna Mylinska, Shutterstock; © sumire8/Shutterstock; © Isuaneye/Shutterstock; © Y.P.photo/Shutterstock
36-37   © Gorodenkoff/Shutterstock; © visiontandel/Shutterstock; © Kamonchai/Shutterstock
40-41   © Marlon Trottmann, Shutterstock; © agefotostock/Alamy Images